Materials in the Environment

Malcolm Dixon

EXPLORING MATERIALS

Building Materials
Materials in the Environment
Materials in your Home
Materials on the Move

Editor: Joanna Housley
Designer: John Christopher

First published in 1993 by Wayland (Publishers) Limited,
61 Western Road, Hove, East Sussex, BN3 1JD, England

British Library Cataloguing in Publication Data

Dixon, Malcolm

   Materials in the Environment. – (Exploring Materials Series)

   I. Title II. Series

   620.1

ISBN 0 7502 0786 8

620.1
M009906

Typeset by Strong Silent Type
Printed and bound in Spain by Graficas Estella.

For Joanne and Michelle

Picture acknowledgements:

The publishers wish to thank the following for supplying the photographs in this book: Cephas Picture Library 30 (Mick Rock); Chapel Studios 24 (Zul Mukhida), 31 (bottom Tim Richardson); Bruce Coleman Limited 18 (Jack S Grove), 39 (John Cancalosi); Eye Ubiquitous 25 (Derek Redfearn); Tony Stone Worldwide 7, 8 (top Alastair Black, bottom Robert Yager), 11 (Christian Bossu-Pica), 12 (Stephen Studd), 19 (Marc Chamberlain), 26 (Martin Rogers), 31 (top, Charlie Waite), 35, 36 (Mike Abrahams), 38 (Keith Wood), 41 (David Woodfall), 42 (top Dennis O'Clair, bottom Tom Raymond); Wayland Picture Library 4, 9, 34, 37 (© Shell Centre); ZEFA 20 (top). Artwork by Peter Bull.

NOTES FOR PARENTS AND TEACHERS

This book will be useful in implementing aspects of the National Curriculum at Key Stages 1, 2 and 3. Within MATERIALS IN THE ENVIRONMENT there are activities and information which are particularly relevant to Science Attainment Targets 1 (Scientific Investigation), 2 (Life and Living Processes) and 3 (Materials and their Properties). MATERIALS IN THE ENVIRONMENT can be developed as a cross-curricular topic involving Science, Geography, History and English.

There are some activities in this book which will require the help of a parent or teacher. The section on places to visit will be useful to parents during weekends and school holidays.

Contents

Materials and the environment

Our environment is everything around us. In our daily lives many of us are fortunate enough to be able to buy bicycles, clothes, sports equipment, books and pens. We travel long and short distances using cars, buses, trains, ships and aircraft. Our homes are made more comfortable by curtains, carpets, furniture, televisions and hi-fi systems. We eat a wide variety of foods, some of which come from other countries. If we are ill we are treated by doctors and if necessary we go to hospitals. At school we use pencils, rulers and science apparatus. We live in buildings which are warm and secure. We are able to visit great buildings which are hundreds of years old

LEFT
The world around us, and
the materials in it, make
up our environment.

and cross rivers using magnificent bridges. It is easy to forget that many different materials are vitally important in all these areas of our everyday lives.

This book will develop awareness and understanding of materials in the environment and show how we are able to use them to benefit our lives. It also shows the value of natural resources, reminding us

that we must not use up materials that could run out. Scientific developments can mean damaging the Earth, and we need to make sure that we look after our planet.

Throughout the book there are activities and experiments, which give opportunities to gain first-hand experience of the properties of a range of materials in the environment.

Planet Earth

Our planet is like a tiny ball in the vastness of space. The Earth is one of nine planets that orbit the Sun and make up the solar system. The Sun is one of billions of stars in the galaxy called the Milky Way. Our galaxy is just one of billions of galaxies in the universe.

For many centuries people believed that the Earth was flat. The Earth has now been photographed many times from outer space and the photographs confirm that our planet is shaped like a round ball. This shape is called a sphere or globe. The Earth measures 12,735 km across its middle.

For thousands of years, it was believed that the centre of the Earth was solid. Even now no-one is really certain what it is like inside. We have only been able to drill into the Earth to a depth of about 14 km. Scientists have, however, been able to build up a picture of the Earth's interior by studying the shock waves from earthquakes. Seismic stations around the world record these waves. The scientists can then work out the types of material that these waves have passed through. Other evidence about the Earth's interior comes from rocks found in the lava from volcanoes. Rocks from space - called meteorites - also provide clues since they were probably formed at the same time as the Earth.

RIGHT The Earth is like a tiny ball whirling around in space.

ABOVE An erupting volcano, in Hawaii, releases red hot lava, cinder and gases.

The evidence suggests that our planet has a layered structure. We live on the outer layer, called the crust, which is made of solid rock. The solid crust varies in thickness from only 6 km, under the oceans, to 90 km under the land. Beneath the crust is the mantle. The mantle is 3,000 km thick and is made of extremely hot rock containing mainly silicon, magnesium and oxygen. The rocks here are so hot that they are as soft as treacle. Sometimes these molten (melted) rocks, or magma, emerge through undersea volcanoes. These hot rocks then solidify to form underwater mountains.

The mantle covers the core, which is divided into inner and outer sections. The outer core is even hotter than the mantle. It contains molten nickel and iron. The inner core is solid nickel and iron. The temperature here is over 3,500° C.

The Earth's crust is divided into sections called tectonic plates. The continents sit on some of these plates. Over millions of years the plates slowly change their positions. It is the movement of these plates against one another that causes earthquakes and volcanic eruptions.

Air around us

We live on planet Earth surrounded by air. On windy days we feel the air blowing on our faces and hear it as it rushes between buildings. We see leaves blown along by it, but we cannot smell or taste it. Air is transparent and for much of the time we forget about it; yet air is vital for life on Earth.

This ocean of air - called the atmosphere - consists mainly (78 per cent) of a gas called nitrogen. Most of the rest (21 per cent) is made up of the important gas oxygen. Oxygen is needed for burning and by all living things for breathing. The rest of the atmosphere is made up of other gases such as argon, neon, helium, krypton, ozone, hydrogen, carbon dioxide and water vapour. Many of these gases can be separated from air and are then used by humans. Nitrogen is used, in liquid form, for freezing food. Oxygen is used in steel-working and welding. It is also essential in hospitals to help patients with breathing difficulties.

The atmosphere is made up of very small particles, or molecules, which are so small they cannot be seen. Since air is made up of a mixture of gases, then these particles are also of several different kinds. Each of these particles has mass and is, therefore, pulled towards Earth by the force of gravity. All of the air around us is pulled down towards the Earth by gravity.

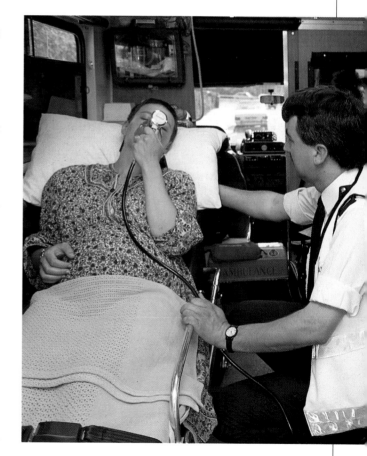

RIGHT This woman is being given oxygen to help her breathe.

RIGHT
Aeroplanes fly in the layer
of the atmosphere called
the troposphere, which is
closest to Earth.

The atmosphere can be divided into five layers. It is only in the layer closest to the Earth, called the troposphere, that there is enough oxygen for us to breathe. In fact, most of the gases of the atmosphere (about 80 per cent) are contained within this layer, which extends up to 15 km above the Earth's surface. It is here that clouds form and weather conditions develop. All animal flight and most aircraft flight takes place in this layer. The next layer is called the stratosphere. It extends up 50 km and contains 19 per cent of the atmosphere's gases. The three remaining layers are the mesosphere, ionosphere and exosphere. Beyond this is the air-less emptiness we call space.

The ocean of air presses down on everything in it. This force is called atmospheric pressure. It was first discovered by an Italian scientist, Evangelista Torricelli, in 1641. On the ground the pressure of air is high. The air presses in all directions - up, down and sideways - as the molecules in it zoom about at terrific speed. For example, when a rubber plunger is pressed on a surface, the air is squeezed out from underneath it. The pressure on the plunger from the outside is greater than that inside. The plunger has to be pulled extremely hard to lift it up. This gives an idea of the strength of air pressure. Air pressure decreases as we climb higher into the atmosphere.

We need air from the atmosphere in order to live. We breathe air in through our noses and it travels down a tube, called the windpipe, into our lungs. In our lungs oxygen, from the air, passes into blood vessels and is taken to every part of our body. When we breathe out, the gases which are expelled include oxygen, nitrogen and carbon dioxide. When we move out of normal air pressure and oxygen supply we have to provide our own oxygen. For example, mountain climbers sometimes have to carry their own oxygen air cylinders. In space, there is no air so astronauts need to take oxygen with them.

Even though the atmosphere is so important to us, humans have allowed their many activities and inventions to affect it. Car engines, power stations, domestic fires, industry, cigarette smoking and many other things release substances into the air causing it to become polluted. Coal and oil-fired power stations produce gases, such as sulphur dioxide, which have made the rain more acidic. Acid rain has destroyed forests and wildlife around the world.

BELOW The air is polluted by smoke from industrial plants.

Heat from the Sun is trapped by gases in the atmosphere, including methane and carbon dioxide, so ensuring that the planet is kept warm. This is known as the greenhouse effect. Scientists have found that these 'greenhouse gases' are increasing, because of human activity, and this could cause the planet to get warmer. The result might be more hurricanes, floods and famine. Some gases produced and used on Earth, such as chlorofluorocarbons (CFCs), found in aerosols and refrigerators, are now thought to be damaging part of the atmosphere known as the ozone layer. Ozone is a gas found high in the atmosphere. It absorbs some of the harmful rays from the Sun, and so protects us from serious illness.

LEFT
These trees in eastern Europe have been damaged by acid rain.

Oxygen and burning

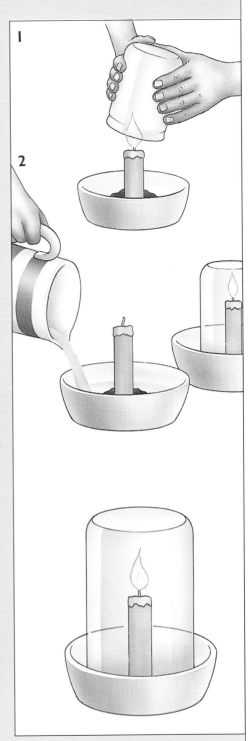

1. Fix the base of one candle in Plasticine. Place it in the centre of one bowl. Ask an adult to light the candle. Place the smaller of the glass jars over the candle. Watch carefully to see what happens.

2. Use Plasticine to fix the other candle in the centre of the second bowl. Pour some water into this bowl. Ask an adult to light the candle. Place the larger jar over the burning candle. As the candle burns, watch to see what happens to the water level. What happens to the candle flame? Repeat this investigation. Does the same thing happen?

Air contains the gas oxygen which is needed for burning. When the candle flames go out, all the oxygen in the jars has been used up. Did the candle burn longer in the larger jar? Why did this happen? As the oxygen is used up, the water rises up the jar to fill the space it leaves. Can you tell from your investigation what fraction of the air in the jar is oxygen?

Air has mass

1. Blow up both balloons. Tie thread around the necks of the balloons so the air cannot escape. Use thread to tie each balloon to the ends of a length of dowel.

2. Balance the dowel rod across a pencil resting on two boxes as shown. The balloons should now be level. Let the air out of one balloon. What happens? The air in the balloon is made up of matter and therefore it has mass. The air in a large room has about the same mass as an adult.

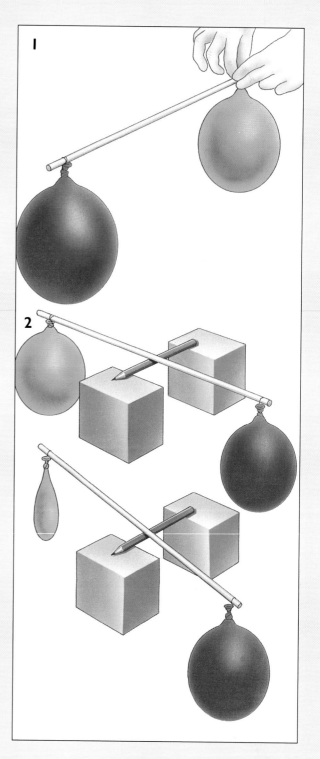

Air pushes

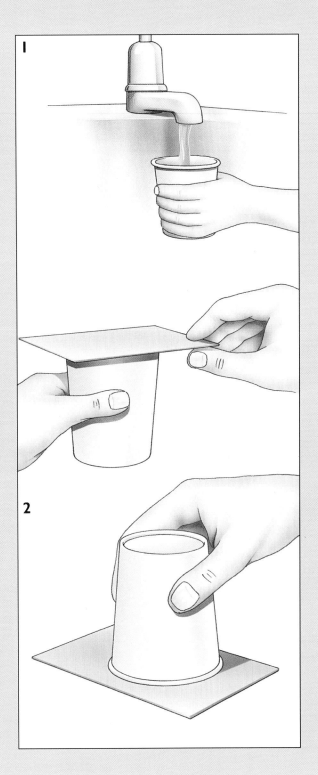

YOU WILL NEED

a plastic beaker
a jug of water
stiff card (a postcard is suitable)
a large plastic bowl

1. Fill the beaker to the top with water. Place the card over the top of the beaker. Stand it over a large bowl or sink.

2. Hold the card over the beaker and turn the whole lot upside down. Carefully take your hand away. What happens? Air is pushing against the card and this keeps the water in the beaker.

Using air pressure

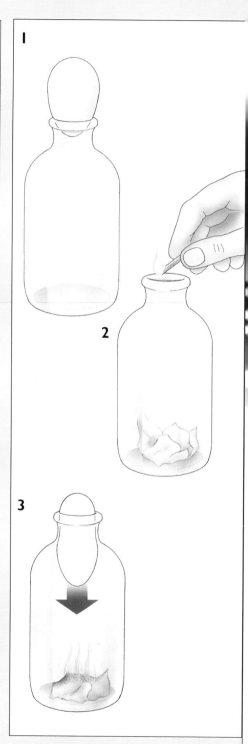

1. Remove the shell from the egg. Place the egg
   on the neck of the bottle. It should fit into the
   neck of the bottle but be unable to fall
   through.

2. Remove the egg. Crumple up a piece of paper
   and place it in the bottle. Ask an adult to light
   the paper by dropping a burning match into the
   bottle.

3. Quickly place the egg on the neck of the
   bottle. Watch what happens. How can you
   explain this?

The paper burns inside the bottle, using up the
oxygen and reducing the air pressure inside the
jar. The air pressure outside the jar is greater
than that inside it. Air presses on the egg,
forcing it into the jar.

Measure the wind speed

<u>YOU WILL NEED</u>

a strong cardboard box   scissors
dowel rod   strong glue
a stapler

1. Cut away both ends of the cardboard box. Punch holes near one end as shown. Cut a length of dowel rod to fit through these holes.

2. Cut out a strong cardboard flap from one of the ends, and glue or staple this to the dowel rod. The flap must be able to swing freely.

3. Cut out an arrow shape and glue this to one end of the dowel rod. When the flap moves the arrow should also move. Make a curved scale from 1 to 10 on the side of the box.

4. Hold your wind-speed measurer so that the flap faces the wind. Does the flap move? What number does the arrow move to? Measure the wind speed every day for two weeks. Record your results and use them to plot a graph.

In 1806 an Englishman, Admiral Sir Francis Beaufort, introduced a way of comparing wind strengths. His Beaufort Scale is still used today. Look at reference books to find out more about this scale. Listen to weather and shipping forecasts on the radio and television.

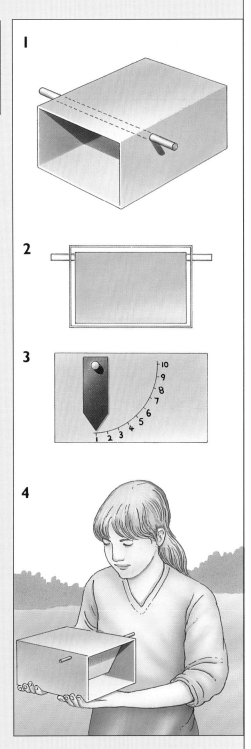

Water

About 70 per cent of the Earth's surface is covered by water in the form of oceans, rivers and seas. The Pacific Ocean covers an area which is larger than all the Earth's land added together.

Living things need water. In fact, most living things are made largely of water. Most of a jellyfish's mass is water. Your body is about 65 per cent water. More than half our food is water. We drink water and use it for cooking, washing up, keeping ourselves clean and for cleaning our homes and cars. Water is vital to all life on Earth. Water is also used in industry, in factories and power stations. More than 50,000 litres of water are used in the manufacture of a car. A large power station uses 5 million litres of water every day for cooling.

Water is a common but remarkable substance. It is unusual because it occurs in all three forms of matter as a liquid, a solid (ice) and as a gas (water vapour). Water freezes at $0°$ C and boils at $100°$ C. So water is a liquid in most places on the Earth's surface. About 2 per cent of the world's water is found as solid ice, for example at the North and South Poles. An extremely small, but very important, part of the world's water occurs in the form of a gas (water vapour) in the atmosphere.

LEFT
Huge stretches of ice, called glaciers, are found in Antarctica.

ABOVE Thousands of plants and aquatic animals live in the world's oceans.

An important property of water is that it can dissolve (make into a liquid) many other materials. Sea water is salty. The salt has come from small amounts of rock that have dissolved in water over millions of years. Water also dissolves air. This means that fish and other aquatic life can breathe under water in seas and rivers.

Providing a supply of clean water for people to use is a big challenge for engineers. A large city needs over 500,000 million litres of water, free from harmful impurities, every day. Some water, especially in the Middle East, is obtained from sea-water. The sea-water is passed through a desalination plant which removes the salt and leaves behind almost pure water. A less expensive method is to use water from lakes, rivers and underground wells. The water is cleaned using filter beds. These beds contain sand and gravel which filter out unwanted impurities. The water is then stored in reservoirs. Chlorine is added to the water to kill any remaining bacteria. Careful testing takes place to make sure that clean water reaches our homes.

RIGHT
Special water treatment
plants are used to purify
the water that we use in
our homes.

In some areas the water contains a material called fluoride, even after filtering. Dentists discovered that fluoride helped children's teeth to grow strong and resist decay. As a result fluoride is added, in small amounts, to the water in many areas where it is not found naturally.

After it is used, and dirt and waste materials are added to it, the water becomes sewage. Sewage, from the drains, is pumped into large tanks which allow solid waste materials to sink to the bottom. The remaining liquid part is then filtered and treated with chlorine. Then, because it is almost pure, it can be allowed to flow back into rivers.

The water on Earth is constantly being recycled. Water evaporates from the earth's surface creating water vapour in the air. The water vapour condenses to form clouds. This water returns to the surface as rain, snow and hail. This process is known as the water cycle.

Dissolving in water

some clear plastic pots   sugar
a teaspoon   salt
a wide, shallow,   flour
coloured dish   curry powder
a jug of water

1. Wash the clear plastic pots to make sure they are clean. Pour some cold water into one pot. Add a spoonful of salt to the water. Stir until the salt disappears. Scientists say that the salt has dissolved in the water. The salt is called the solute and the water is the solvent. The mixture of water and salt is called a solution.

2. When there is one teaspoon of salt in a small pot of water the solution is said to be dilute. Add more salt to the water. Keep stirring. The solution is getting more and more concentrated. When no more salt will dissolve, the solution is said to be saturated.

3. Pour the saturated solution into a wide coloured dish. Watch what happens over a few days. What is left in the dish?

4. Try dissolving sugar, flour, curry powder and other kitchen materials in water. Do they all dissolve? Try using warm water. Those which do not dissolve are called insoluble.

Investigate some properties of water

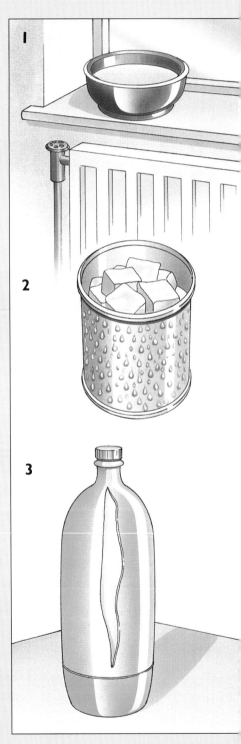

1. Leave a small bowl of water in a warm place for a few days. What do you think will happen to the water? We say that the water has evaporated. Liquid water has changed into water vapour. Water vapour is an invisible gas. Try leaving small amounts of water in different places: somewhere warm and somewhere cold. Where does evaporation take place fastest?

2. Fill a metal can with ice cubes. Look carefully at the outside of the can. What can you see? Where does it come from? Water vapour has condensed on the cold surface of the can into small liquid water droplets.

   Look at your bedroom windows on a cold morning. What can you see on the glass?

3. Fill a plastic bottle to the top with water. Screw on the top tightly. Leave it in a freezer overnight. What has happened? Does this help you to understand why water pipes can split in cold weather?

Purify water

YOU WILL NEED

a large, clear plastic bottle
scissors
a bucket of dirty water
(add soil and leaves to tap water)
cotton wool
washed fine gravel
washed coarse gravel
clean sand

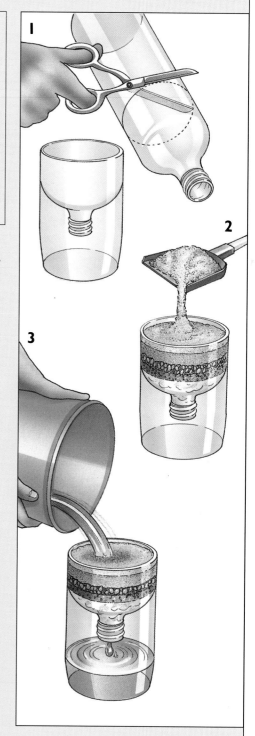

1. Remove the screw top from the bottle. Cut off the top section of the plastic bottle. Fit the top part into the base section as shown.

2. Place some cotton wool in the neck of the bottle. On top of this add a layer of clean, coarse gravel. This gravel layer should be about 5 cm thick. Add a second layer of fine gravel. Now add a 10 cm layer of clean sand.

3. Carefully pour some dirty water from the bucket on to the sand. Watch carefully. Is the dirt being separated from the water? Is the water in the base of the bottle much cleaner than the water in the bucket? Can you think of a way to improve the filtering? Write to your local water company to find out where the water in your area is purified.

Rocks and minerals

The Earth is covered in rocks. They can be found in many shapes, sizes and colours. The Earth's rocks can be divided into three types: igneous, sedimentary and metamorphic. Igneous rocks have solidified from molten rock or magma. Nearly all igneous rocks are made of crystals which interlock. If you examine an igneous rock, such as granite, you will be able to see the different crystals.

BELOW
You can see different layers of rock in this cliff face.

ABOVE A well-preserved fossil found in Dorset, England.

Sedimentary rocks are formed from fragments of older rocks or from the remains of animals or plants. The weather, rivers, ice and seas break up older rocks into fragments. Some of these fragments are pebble size and others are too fine to see. Over millions of years the fragments form layers which are squeezed together, producing sedimentary rock. Sandstone is a sedimentary rock formed when layers of sand are pressed together. Limestone is a sedimentary rock made from the shells and skeletons of animals. Coal is made from the remains of plants which covered parts of the Earth millions of years ago. These horizontal layers (called strata) of sedimentary rock can sometimes be seen in cliff faces. Animals and plants that lived millions of years ago can be found, in the form of fossils, within layers of sedimentary rock.

Metamorphic rocks have been changed or metamorphosed. This change may happen because of the heat of a volcano or because of the pressure from other rocks. Limestone is metamorphosed into marble. Slate is another common metamorphic rock. It is formed by squeezing clay.

All rocks are made up of minerals. A mineral is a natural substance. There are thousands of different minerals. Some rocks are made up of one mineral but most are mixtures of minerals. Scientists recognize and identify different minerals by their properties. They will observe the shape and colour of the minerals. They might test the hardness and see if the minerals react with acid. More complicated tests will be carried out in their laboratories. Like all materials, minerals are themselves made of the 'building-blocks' called the elements. Two of these elements, oxygen and silicon, are extremely common amongst minerals.

Some minerals are said to be precious and are valued for their beauty. Diamonds, rubies, sapphires and emeralds are amongst this group. They are carefully cut and polished before being made into jewellery. Diamonds are made by intense heat and pressure deep inside the Earth. One property of diamond is that it is the hardest of all natural substances.

RIGHT Diamonds can be cut, polished and made into sparkling jewellery.

Important metals, such as aluminium and copper, are found in some minerals. These minerals are often mixed with other substances to form ores. These ores are distributed throughout the world. Rocks and minerals are very important as the raw materials we use in manufacturing.

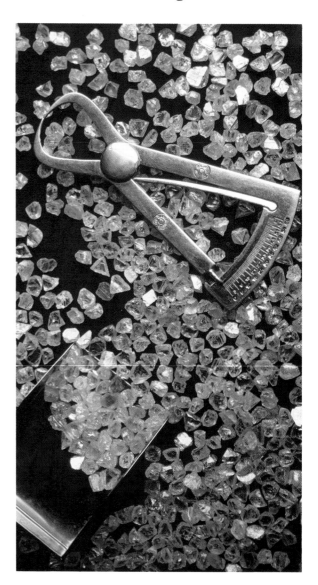

Collect rocks

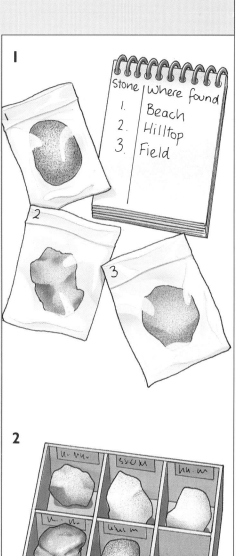

1. Build up a rock collection by gathering small rocks from different places. Ask an adult to help you if you want to collect from dangerous places. Start by looking in your garden and school grounds. Try to collect some samples from dried-up river beds, the seashore and old quarries. Look for small loose stones with interesting shapes, colours and patterns. Write down, in your notebook, the place where you found each rock. Place each stone, and your note of where it was found, in a small plastic bag.

2. Wash your rocks and let them dry. Store them in cardboard boxes. Use card to make divisions in each box, as shown. Use reference books to try to identify each rock. Write a label for each one showing the type of rock it is, and where you found it.

3. Visit museums to find out more about the rocks in your collection.

Investigating rocks

a piece of granite (igneous rock)
pieces of limestone and chalk
(sedimentary rocks)
pieces of marble and slate
(metamorphic rocks)
a variety of other rocks

vinegar (acetic acid)
a dropper
a magnifying lens
a copper coin
a sharp knife

1. Use the magnifying lens to look at the piece of
   granite. Can you see quartz (white), mica
   (black) and feldspar (pink) crystals in it?

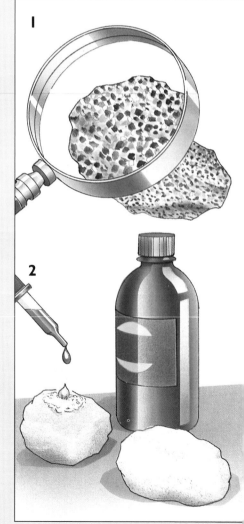

2. Drop some vinegar on to your sample of
   limestone. Use the magnifying lens to see what
   happens. Is there any fizzing? Test the chalk in
   the same way. Many rocks contain a mineral
   called calcite (calcium carbonate). When an
   acid is mixed with calcite it gives off a gas called
   carbon dioxide. Vinegar is a weak acid. The
   fizzing that you see is carbon dioxide being
   produced. Do limestone and chalk contain
   calcite?

3. Use your magnifying lens to examine the
   sample of slate. Can you see that it is in layers?
   Examine the marble. In what ways is it similar
   to slate? What differences can you see?
   Test the slate and marble and granite by
   dropping vinegar on to them. Do they contain
   calcite?

3

4

5

4. Examine each of the rock samples with a magnifying lens. Make a note of the colours that you see on each one. Are any of the rocks see-through? Are they opaque? Are they shiny? Do they feel rough or smooth? Can you see any crystals? Can you see small stones or sand? Write a few sentences to describe each rock.

5. Work out which of your rocks is the hardest. Start by scratching each rock with your fingernail. Does it leave a mark on some rocks? Test the rest with a copper coin and then ask an adult to test them with the blade of a sharp knife. Try to put your rocks in order of hardness.

In 1812 a German geologist called Friedrich Mohs worked out a scale of hardness for minerals from diamond (10) to talc (1). Diamond is the hardest of all minerals and can scratch all other minerals. Which minerals, on the scale, will quartz scratch?

Diamond	10
Corundum	9
Topaz	8
Quartz	7
Feldspar	6
Apatite	5
Fluorspar	4
Calcite	3
Gypsum	2
Talc	1

Soil

Soil is one of the most valuable materials found on the Earth. We eat plants which grow in the soil. We use trees, which also grow in the soil, to provide us with timber and paper. We obtain meat, milk and different forms of clothing from animals. These animals feed on plants growing in the soil.

On land there are some mountainous areas with no soil covering the rocks. There are other places with poor soils that allow few plants to grow. It is important, especially as the population of the world is increasing, that we look after those areas of land where there are fertile soils.

Soil is formed from the solid rocks on the Earth's surface. These rocks have been, and are being, broken down by weathering and erosion. Frost is an important method of erosion. Most rocks contain water. When the water freezes it can produce enough pressure to shatter the rocks.

Heavy rain and the direct heat of the sun cause the surfaces of rocks to break up. The action of fast-moving seas, rivers, streams and winds makes rock material wear away. Small plants, such as lichens and mosses, grow amongst the smaller pieces of rock. When these plants die and decay their material

LEFT
We obtain meat and wool from sheep. They feed on plants growing in soil.

LEFT
Rain water flowing down
hillsides cause soil erosion.

is then added to the particles of rock. The soil is now able to support larger plants. In time their dead leaves, stems and roots are added to the soil. Tiny organisms, such as bacteria and fungi, and larger creatures, such as earthworms and woodlice, break the soil down further. Earthworms are important because they swallow soil, allow it to pass through their bodies, and expel it in the form of worm casts. They mix the soil up and break it into fine pieces. Their burrows allow air and water to get into the soil. They drag dead leaves into the soil where they decay.

There are many different types of soil. The type of soil in any area is affected by the climate and the materials from which the soil has developed. When you look carefully at a sample of soil, you can see that it is a mixture of particles. Some of the particles are rock, others are the remains of dead animals and plants.

RIGHT
Earthworms are amongst the most
important animals in the soil.

Investigate soil

_____ YOU WILL NEED _____

an old spoon
samples of soil from different places,
such as a garden, waste ground,
a wood, a field

sheets of white paper
some clear glass or plastic jars with lids
a jug of water
a magnifying lens

● **WEAR GLOVES WHEN WORKING WITH SOIL.**
WASH YOUR HANDS THOROUGHLY AFTER HANDLING SOIL.

1. Place some soil from each place on sheets of white paper. What differences are there between the soil samples? Do they feel different? Use the magnifying lens to look at the soil samples. Do they smell differently? Write down your observations.

2. Put a sample of one type of soil in a clean, clear jar. Pour in water until the jar is about three-quarters full. Fix the lid on the jar and shake it well for about a minute. Leave the jar on a window sill. Watch how the soil settles. Can you see layers of different sized particles? Which layer has the smallest particles? Which layer has the largest particles?

3. Do the same thing with soil samples from other places. Do they separate into layers? How are the layers in each sample different? In what ways are they alike? Try to collect soil samples from different areas. Repeat your tests with these samples.

Dig a soil profile

_____ YOU WILL NEED _____

a gardening spade
a metre ruler

● **YOU WILL NEED AN ADULT
TO HELP YOU**

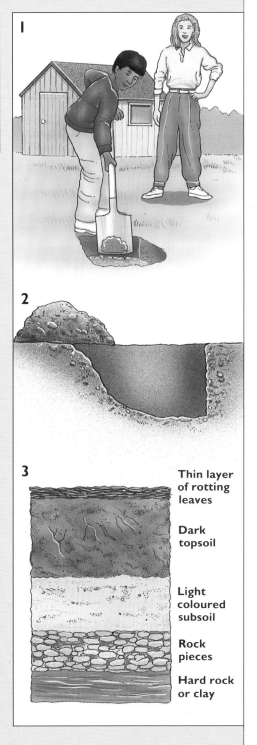

Thin layer
of rotting
leaves

Dark
topsoil

Light
coloured
subsoil

Rock
pieces

Hard rock
or clay

1. Ask an adult to help you find a piece of ground where you can dig a deep hole. It needs to be a safe, undisturbed place in the corner of a garden or field, where you will not dig up any plants.

2. Dig through the soil to 1 m below the surface. Try to keep one side of your hole vertical.

3. Look carefully at the vertical side of your hole. This is known as a soil profile. Can you see different layers? Are they different colours? Measure the depth of each layer. You may be able to see a thin top layer made from rotting leaves. The layer underneath this, called topsoil, is dark, rich in humus and full of worms and other creatures. Below this is the subsoil. It contains very little humus. Underneath this are weathered rock pieces and then a layer of hard rock or clay. Can you see any plant roots? How far down do they go? Remember to replace the soil into the hole when you have finished.

Using raw materials

LEFT Modern chemical plants turn raw materials into new materials which we use every day.

Look around you and you will see many things that have been made from raw materials. We obtain our raw materials from the Earth, from the air, from water around us and from living things. We use minerals, rocks, fuels, gases, water, animal and plant materials from our environment to make things which we use every day. Raw materials are made into useful products in factories and chemical plants.

The chemical industry started when humans first tried to change the natural materials they found in their environment. Some of the earliest procedures, such as dyeing and making soap, were carried out at low temperatures. Later, higher temperatures were needed to make glass and for extracting metals from their ores.

In 1650, a German chemist called Glauber showed how wood ash could be used to make an artificial fertilizer. In 1740, scientists began to produce sulphuric acid in large quantities. This acid, and others, led to the production of useful gases. In 1779, Charles Tennant found a way to make bleaching powder and this was used, in great quantities, by the growing British cotton industry. In the 1850s, the first steel-making plant was built and, in the USA, the petroleum industry started with the drilling of the first oil well. By this time the explosives, fertilizer and dyeing industries were being established. Since then the chemical industry has changed at a rapid pace to help make our lives more convenient.

Limestone is an important raw material. It has been used in constructing many buildings, and to provide the foundations for roads. It is found near the surface of the Earth and so is quite cheap to extract by quarrying. Millions of tonnes of limestone are quarried every year. Unfortunately, quarrying does affect the environment - the landscape is changed and there is increased traffic, noise and air pollution in the area.

Chemists identify limestone as a form of calcium carbonate and give it the formula $CaCO_3$. All acids, even weak acids such as rainwater, react with carbonates. So limestone in the environment is slowly dissolved by rain. This is why we find some spectacular scenery in limestone areas.

LEFT
Limestone is a raw material which is found just below the Earth's surface.

Millions of tonnes of limestone are used every year to make steel, glass, cement and other chemicals. These materials are vital in our modern world. Without limestone it would be impossible to make them. Limestone also has an important role to play in reducing air pollution from some power stations. It can neutralize the acidic gases that they produce and form a material, calcium sulphate, which can then be used to make building plaster.

Coal is a hard, black, solid material which people have been burning for thousands of years. Some people burn coal on fires in their homes. Most coal is used in power stations to generate electricity. Coke, coal-gas, ammonia and coal-tar can also be obtained from coal. Coke is used in steel-making. Coal-gas and coal-tar were important materials in the chemical industry in the nineteenth century. Ammonia is used to make fertilizers.

BELOW A coalminer digging out coal from a narrow seam.

Coal comes from dead plant material, squashed into seams of rock by layers of sand and mud. This happened during the development of the Earth's crust about 300 million years ago. Coal is made of carbon, hydrogen and oxygen. One form of coal, called anthracite, is almost all carbon. It burns extremely well.

Many seams of coal are found deep in the Earth. The coal is obtained by underground mining. Digging out coal has always been a dangerous job for the miners involved. The coal needs to be cut from the seams, which in a modern coal-mine may be miles under the sea, and then brought to the surface. Coal-mines are now much safer than in the past. Other coal deposits are nearer the surface and are dug out of huge quarries. This is known as open-cast mining.

Billions of tonnes of coal are taken from the Earth every year in countries such as the USA, Russia, China, Australia and the United Kingdom. Coal is an important fuel but burning it causes serious air pollution.

ABOVE The plastic used in all these objects is made from oil.

Crude oil or petroleum is, like coal, an important raw material produced from things that were once living. Crude oil was formed beneath the Earth's surface, millions of years ago, by bacteria acting on plant matter and the remains of marine animals. From crude oil we obtain fuels for transport, factories and power stations. In addition, this oil is the source of thousands of other products that we use every day. Plastics, medicines, paints, detergents, some types of clothing and fertilizers are all produced from crude oil.

RIGHT
Oil is made into different useful materials at massive refineries, like this one in the USA.

Oil is found in more than 70 countries around the world. The USA and Saudi Arabia are amongst the leading oil suppliers. Rigs drill through rock so that this black liquid can be extracted. New discoveries of crude oil are now mostly from beneath the sea-bed. Special rigs have been built so that drilling can take place even in difficult conditions, such as those in the North Sea.

Once oil has been discovered and extracted, it is made into more useful materials at oil refineries. Here the different substances that make up crude oil are separated by a process known as fractional distillation. Crude oil varies according to where it comes from, but the separation process might produce petrol, diesel oil, kerosene and some gases. The diesel oil provides fuel for lorries and trains. The kerosene provides fuel for heating and for aeroplanes. The gases can be used for bottled gas (propane and butane) and for making chemicals. The petrol is used for motor car fuel. Some of the crude oil is converted using a further process called 'cracking'. It is as a result of this procedure, carried out in large 'Cat Cracker' plants, that scientists can make so many types of plastic.

The most plentiful metal in the Earth's crust is aluminium. Like most metals, it is found in ores. The most important aluminium ore is bauxite. Bauxite is found and dug out of mines in Australia, Jamaica and Brazil. The ore is then transported to countries such as the USA, Canada, Germany and Britain. In these countries a process called electrolysis is used to produce pure aluminium. Manufacturing aluminium in this way uses a great deal of electricity. Aluminium is used to make many items in our homes, for building and in many forms of transport, especially aircraft.

BELOW This copper wire was made in Arizona, USA.

Copper mines are found in the USA, Chile, Zambia and a few other countries. The copper ores contain only a very small amount of the metal copper. The ore is therefore crushed to break it up into small pieces. The crushed rock is mixed with water and a detergent. A froth forms and this contains copper. The froth is collected, dried and roasted to form slabs. Electrolysis is used to produce the final, very pure, form of copper which manufacturers need. Copper has many uses in our homes, especially for electric wiring, and is mixed with other metals to form brasses and bronzes.

You may use salt to sprinkle on your food and you may know that salt is used on roads in winter to melt the ice. But you probably do not know that salt, which chemists call sodium chloride, is the raw material for many other chemicals and items in daily use. Salt is pumped up from mines below the surface of the Earth. It is then used in chemicals to make some food products, medicines, cosmetics, some plastics, paper, soaps and household bleach.

Waste materials

When animals and plants die they produce waste. This waste is broken down by worms, bacteria and fungi so that its chemicals and nutrients are returned to the Earth to be used again. Gardeners use compost heaps to make garden waste rot and so produce humus. This humus can be dug back into the soil. So natural materials are dealt with efficiently by nature's recycling system.

ABOVE Huge mounds of rubbish are stored in landfill sites.

Humans produce great amounts of waste materials. Some of these materials are used again but most are deposited in holes in the ground - landfill sites. Many of the materials made by humans, including plastics and glass, do not decompose easily. These materials, and others which are not re-used or recycled, are polluting the environment, so threatening our health and possibly the future of our planet.

Valuable materials such as metals, glass, plastic, cardboard and paper are deposited in dustbins. When the dustbins are collected, the waste is usually deposited in landfill sites or disposed of by incinerating (burning). Decaying waste can produce poisonous liquids and gases which sometimes seep into streams, rivers and even our water supplies. Some gases, such as methane, can cause explosions. When waste materials are burnt, the fumes that are produced pollute the air. It makes sense to recycle as many materials as possible. In more and more areas there are recycling schemes for paper, glass, steel, aluminium and old clothes. These schemes save valuable raw materials and help protect the environment.

Industry uses raw materials to make goods that people want. Inevitably many different types of industries produce waste materials. Much of this waste is harmless but some is poisonous. Most industrial waste is buried in landfill sites. Many of these sites contain dangerous mixtures of chemicals. Some waste materials are so dangerous that it is not known how to dispose of them safely. Radioactive materials are used in industry, scientific research and in making nuclear energy. Some of the waste materials produced in these industries retain their radioactivity for thousands of years.

ABOVE All kinds of paper can be recycled.

It is important that such waste is stored safely to protect people now and in the future. It has to be stored in strong steel containers on special sites or deep underground.

What we do with waste materials is very important. We must try to cut down on the waste we produce and take care not to pollute our world with long lasting waste materials.

LEFT
Radioactive materials need to be stored safely.

Investigate air pollution

YOU WILL NEED

six small pieces of card	a large sheet of card
scissors	sticky tape or Blu-tak
petroleum jelly	a felt tip pen
a hand lens	

1. Number the cards from 1 to 6. Print your name on each one. Spread a small amount of the jelly over the surface of each card.

2. Take three of the cards and place them outside in places where you think the air might be dirty. If you are going to place any of the cards near a road, or other dangerous place, remember to ask an adult to go with you.

3. Take the three remaining cards and place them outside where you think the air is clean. Make a note of where you have placed each card.

4. Leave all the cards for a few days, then collect them. Look at each one with the hand lens. Is there much difference between the cards? Can you identify any of the materials that are sticking to the jelly?

5. Make a large table of your results on a sheet of card, like the one shown on page 43.

3

4

CARD NUMBER	WHERE PLACED	THINGS STICKING TO THE JELLY	WHERE THE DIRT MIGHT COME FROM
1. ANNA	IN SCHOOL CAR PARK		

What you can do

1. Hold a 'recycled' sale in your school. Items for sale could include toys, books and clothes that you no longer need. The clothes would have to be washed before being put on sale. You could also sell things made from items you would normally throw away, such as pen pots made from cardboard tubes.

2. Print a poster, make a video, or organize an exhibition to tell your friends about the need to recycle materials.

3. Set up an 'action group' with your friends to look after your local environment. Try to encourage recycling in your area. Take photographs, write letters and put up posters to make local people more aware of the need to dispose of waste materials thoughtfully.

4. Write to organizations concerned with caring for the environment. You may think it worthwhile joining one of these groups.

RIGHT You could make a caterpillar to keep things in out of an old eggtray.

5. Try to avoid wasting water and other materials at home. Think of ways you can recycle products which are used at home. Look at the empty plastic containers which are thrown away with the rest of your household rubbish. Can you think of any ways they could be used instead of being thrown out?

Topic Web

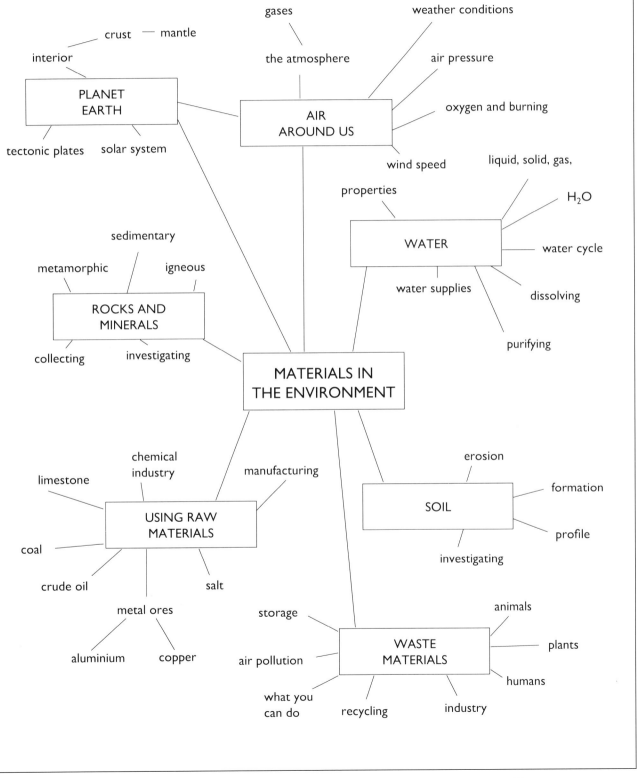

Glossary

Acid A sour substance that can burn.

Ammonia A gas used to make fertilizers and explosives.

Atmosphere The layer of gases surrounding the Earth.

Bacteria Tiny organisms that are present everywhere.

Chlorine A greenish-yellow gas used to purify water.

Decompose To break down dead material, putting nutrients back into the environment.

Desalination Removing salt from sea water.

Distillation Making a liquid pure by heating it to make a gas, then cooling it to make it a liquid again.

Electrolysis A chemical process using electricity.

Element A substance that cannot be split into different substances.

Erosion Wearing down of the land surface by water, ice, wind, waves and weather.

Fertilizer A substance used to help make land produce a lot of crops.

Foundations The base of a building or structure below ground.

Galaxy Millions of stars forming a cluster.

Humus Decomposed plant and animal material that is part of the soil.

Magma Hot molten rock which becomes igneous rock when it cools.

Magnesium A light, white, metallic element.

Molecule The smallest particle of any substance that has the properties of that substance.

Neutralize To make harmless.

Nickel A silvery-white metal.

Nutrients The materials which are necessary for growth and life – such as water, fats, minerals and carbohydrates.

Opaque Not letting light pass through.

Orbit The path of a planet, such as the Earth, around another planet or star, such as the Sun.

Particle A very tiny portion.

Pollution Damage caused to the environment by substances released into it.

Radioactive A radioactive substance is an element which breaks down into another element, at the same time giving out harmful rays, particles or waves.

Seismic stations Places where earthquake shocks are recorded.

Silicon A non-metallic element.

Solar system The group of planets in orbit around the Sun.

Universe Everything - including planets and stars - that exists anywhere.

Welding Joining pieces of metal together by melting and pressing them together.

Books to read

Exploring Soils and Rocks by
E Catherall (Wayland, 1990)

Raw Materials by R Kerrod
(Wayland, 1990)

The Ozone Layer by A Hare
(Gloucester Press, 1990)

This Fragile Earth by J Baines
(Simon and Schuster, 1991)

The Environment by C Twist
(Wayland, 1990)

Experiments with Air by
B Murphy (Franklin Watts,
1991)

Rock Collecting by R Gans
(Black, 1989)

Water by B Walpole
(Kingfisher Books, 1987)

*Earth, the Incredible Recycling
Machine* by P Bennett
(Wayland, 1993)

Manufacturing Industry by
R Kerrod (Wayland, 1990)

Places to visit

Catalyst Museum
Mersey Road
Widnes
Cheshire WA3 0DF

Body Shop Factory
Watersmead
Littlehampton
West Sussex BN17 6LS

Heights of Abraham
Country Park
Matlock
Derbyshire

National Stone Centre
Wirksworth
Derbyshire

The Discovery Centre
77-79 Vyse Street
Hockley
Birmingham B18 6HA

Eureka!
The Museum for Children
Discovery Road
Halifax HX1 2NE

Geological Museum
Exhibition Road
London SW7 2DE

Big Pit Mining Museum
Blaenfon
Gwent
NP4 9XP

Sunderland Museum and Art
Gallery
Fawcett Street
Sunderland
Tyne and Wear

Llechwedd Slate Caverns
Blaenau Ffestiniog
Gwynedd
North Wales

Maritime Museum
Albert Dock
Liverpool

Science Museum
Exhibition Road
London SW7 2DD

Index